PATTERNS IN THE SKY

CLOUDS and Their Patterns

by Thomas K. Adamson

PEBBLE
a capstone imprint

Published by Pebble, an imprint of Capstone.
1710 Roe Crest Drive, North Mankato, Minnesota 56003
capstonepub.com

Library of Congress Cataloging-in-Publication Data is available on the Library of Congress website.
ISBN: 9781666354997 (hardcover)
ISBN: 9781666355031 (paperback)
ISBN: 9781666355079 (ebook PDF)

Summary: What are clouds? How are they made? And why do they seem to move cross the sky? Answer these questions and more and discover the science behind clouds and their patterns.

Editorial Credits
Editor: Alison Deering; Designer: Sarah Bennett; Media Researchers: Julie De Adder and Svetlana Zhurkin; Production Specialist: Katy LaVigne

Image Credits
Shutterstock: AlexReut, 6, Binson Calfort, 10, Calin Tatu, 13, Dark Moon Pictures, 8–9, John D Sirlin, 14–15, kikk, 1, KpaTyH, 5, kuruneko, 17, Piyawat Hirunwattanasuk, 7, Studio-M, 18–19, Sunny Forest, 4 (top left) and throughout, Triff, cover, Vasiliy Merkushev, 11, VisanuPhotoshop, 4 (bottom), Wirestock Creators, 16; Svetlana Zhurkin, 20, 21

All internet sites appearing in back matter were available and accurate when this book was sent to press.

Printed in the United States 6475

Table of Contents

Words in **bold** are in the glossary.

What Are Clouds?

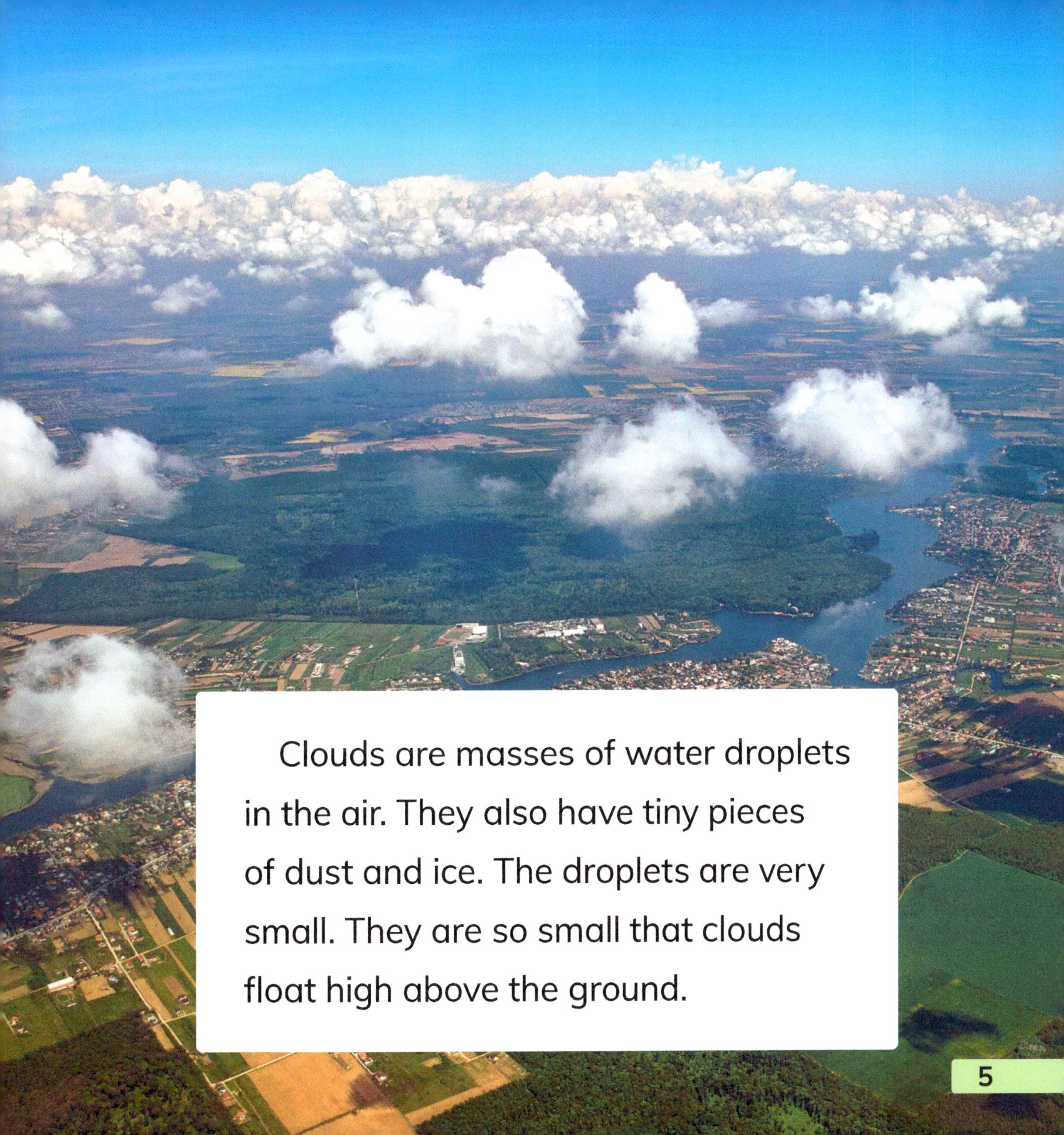

Clouds are masses of water droplets in the air. They also have tiny pieces of dust and ice. The droplets are very small. They are so small that clouds float high above the ground.

How Do Clouds Form?

The air around us is full of water we can't see! Water **evaporates** from Earth's surface. It turns into **water vapor** in the air.

Warm air near the ground rises and cools. The water vapor **condenses** onto tiny dust pieces. Those water droplets grow larger. They come together to form a cloud.

Where Does Rain Come From?

The water droplets in a cloud can grow bigger. They might become heavy enough to fall to the ground. That's rain!

Clouds are high enough in the sky that water can freeze into tiny ice pieces. Those pieces can fall as hail, sleet, or snow. They are all different kinds of **precipitation**.

How Do Clouds Move?

The wind pushes clouds along. It can change their shape. Clouds also move water from place to place.

Clouds are a key part of the **water cycle**. Water evaporates from the ocean and the land. It condenses into clouds. The clouds bring rain to different places.

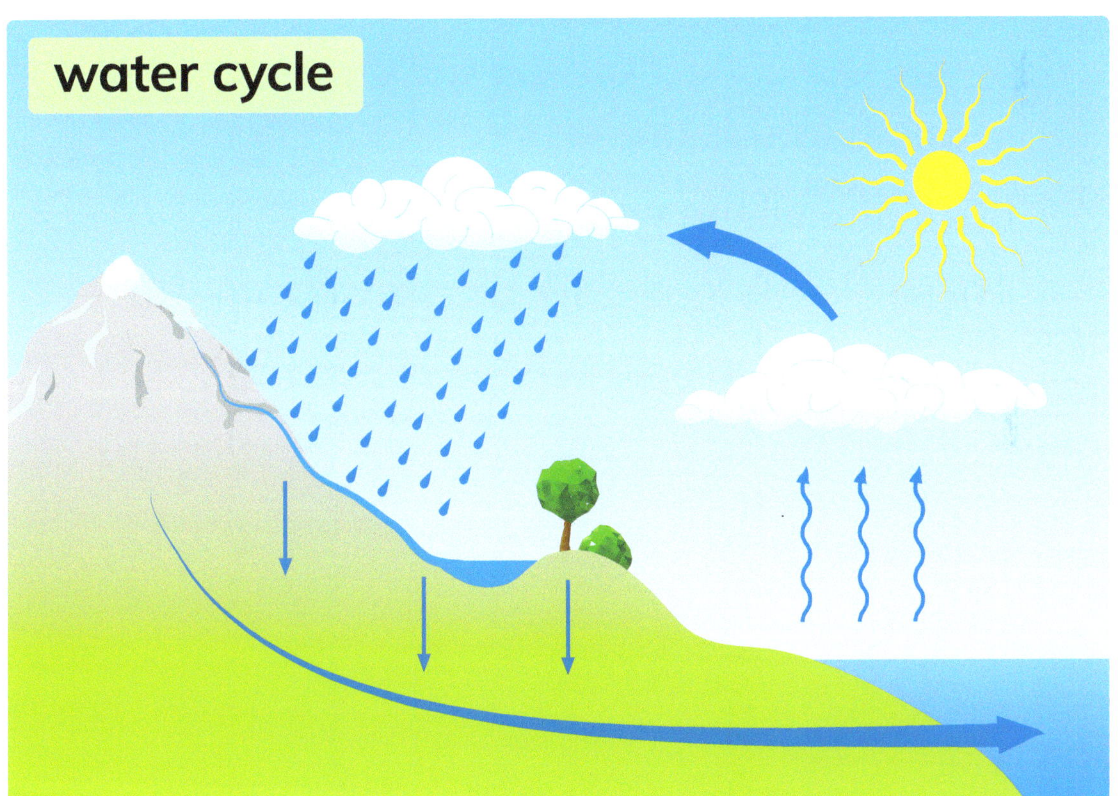

What Color Are Clouds?

Clouds are white when sunlight bounces off them. The light scatters in all the colors of the rainbow. Those colors combine to look white.

Some clouds are very tall. Light does not shine all the way through them. These clouds look dark on the bottom.

Do Clouds Make Noise?

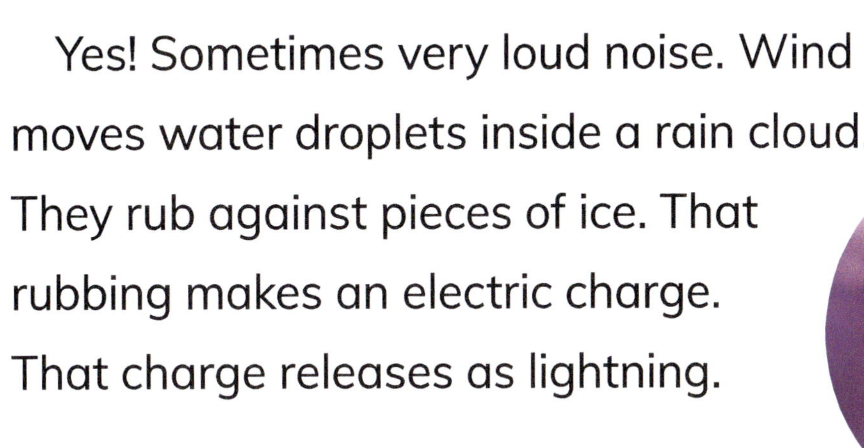

Yes! Sometimes very loud noise. Wind moves water droplets inside a rain cloud. They rub against pieces of ice. That rubbing makes an electric charge. That charge releases as lightning.

Thunder is the sound of the lightning. Lightning heats the air around it. In a second, the heated air expands. The air makes loud crackling and rumbling sounds.

Why Are Cloudy Days Cooler than Sunny Days?

Clouds can block the sun's light. That makes the temperature on the ground cooler.

Clouds can also trap the sun's heat. High, thin clouds trap more heat. Low, thick clouds mostly reflect the sun's heat.

What Are the Different Types of Clouds?

There are many different types of clouds. **Cumulus** clouds look fluffy. **Stratus** clouds are thin and white and cover much of the sky.

cumulus clouds

stratus clouds

Cirrus clouds are thin and wispy. **Altocumulus** clouds are patchy white and gray. **Cumulonimbus** clouds are very tall and cause thunderstorms.

cumulonimbus clouds

cirrus clouds

altocumulus clouds

Water Condensation

Try this simple activity to see water condensation in action!

What You Need

- 2 clear plastic containers with lids
- warm water
- 2 resealable plastic bags
- ice

What You Do

1. Fill one plastic container about a third full with warm water. Leave the other container empty.
2. Place lids on both containers.
3. Fill the two resealable bags with ice.
4. Place a bag of ice on top of each container.
5. Watch what happens. After a few minutes, the sides of the container filled with warm water will get foggy as water condenses. The water vapor rises. When it meets the cold air under the lid, the water vapor cools and condenses. You might even see drops of water form on the sides of that container.
6. Remove the ice and open the lid. See how wet the inside of the lid got? Now take the lid off the empty container. How does that one compare?

Glossary

altocumulus (ahl-toh-KYOO-myuh-lus)—high, layered clouds with gray and white ripples

cirrus (SEER-uhs)—high, thin clouds made of ice crystals that look like strands of white silk

condense (kuhn-DENS)—to change from gas to liquid; water vapor condenses into liquid water

cumulonimbus (kyoo-myuh-loh-NIM-bus)—a huge cloud that can bring hail and tornadoes

cumulus (KYOO-myuh-luhs)—a white, puffy cloud with a flat, rounded base

evaporate (ih-VA-puh-rayt)—to change from a liquid to a gas

precipitation (pri-sip-ih-TAY-shuhn)—water that falls from clouds in the form of rain, hail, sleet, or snow

stratus (STRA-tuhss)—a low cloud that forms over a large area; stratus clouds often bring light rain

water cycle (WAH-tur SY-kuhl)—how water changes as it travels around the world and between the ground and the air

water vapor (WAH-tur VAY-pur)—water in gas form; water vapor is one of many invisible gases in air

Read More

Carlson-Berne, Emma. *Let's Explore the Water Cycle.* Minneapolis: Lerner Publications, 2021.

O'Brien, Cynthia. *Thunder and Lightning Storms: Causes and Effects.* Mankato, MN: 12-Story Library, 2021.

Sohn, Emily, and Erin Ash Sullivan. *Weather and the Water Cycle.* Chicago: Norwood House Press, 2020.

Internet Sites

Weather WizKids: Clouds
weatherwizkids.com/?page_id=64

Easy Science for Kids: Different Types of Clouds
easyscienceforkids.com/all-about-clouds

SciJinks: What Makes It Rain?
scijinks.gov/rain

Index

About the Author

Thomas K. Adamson has written lots of nonfiction books for kids. Sports, math, science, cool vehicles—a little of everything! When not writing, he likes to hike, watch movies, eat pizza, and of course, read. Tom lives in South Dakota with his wife, two sons, and a Morkie named Moe.